W9-BNR-139

CITRUS FRUITS

Susan Wake

Illustrations by John Yates

Carolrhoda Books, Inc./Minneapolis

All words that appear in **bold** are explained in the glossary on page 30

First published in the U.S. in 1990 by Carolrhoda Books, Inc.
 First published 1989 by Wayland (Publishers) Ltd.

Library of Congress Cataloging-in-Publication Data

Wake, Susan
Citrus fruits / Susan Wake ; illustrations by John Yates.
p. cm.—(Foods we eat)
Includes index.
Summary: Describes several citrus fruits, their importance, and their histories and presents several recipes.
ISBN 0-87614-389-3 (lib. bdg.)
1. Citrus fruits—Juvenile literature. 2. Cookery (Citrus fruits)—Juvenile literature. [1. Citrus fruits.] I. Yates, John, ill. II. Title. III. Series: Foods we eat (Minneapolis, Minn.)
SB369.W25 1990
634′.304—dc20
89-32162
CIP
AC

Printed in Italy by G. Canale C.S.p.A., Turin
Bound in the United States of America

Contents

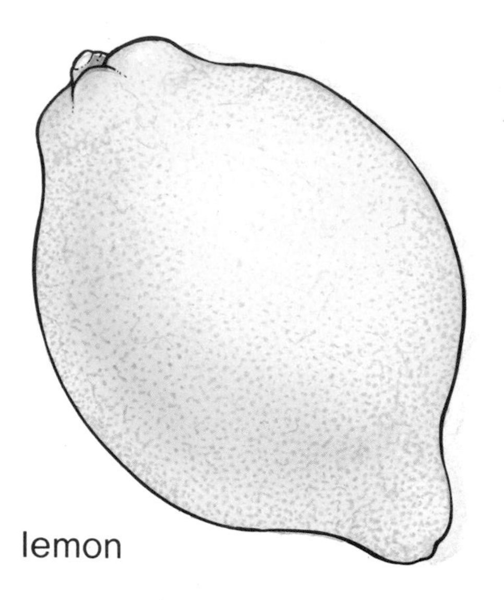

lemon

What are citrus fruits?

A fruit is the edible part of a plant that contains the seeds. Fruit has been an important part of human diets for thousands of years. The first humans ate the fruit and roots of wild plants and later grew them for food. As people migrated to different parts of the world, they brought their seeds and plants with them.

There are many different types of fruit. The

orange

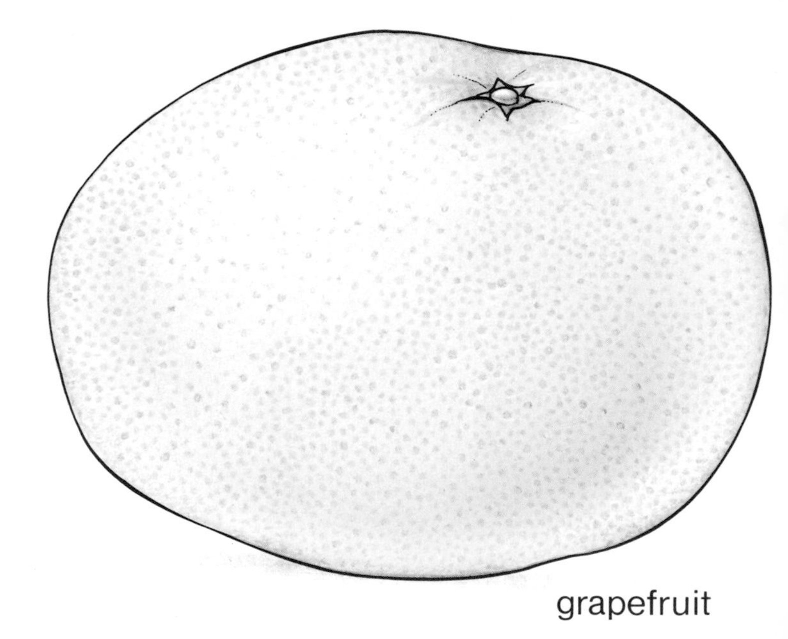

grapefruit

different types can be divided into groups called **families**. Some fruits belong to the **berry** family. Berries are fruits that have their seeds in their **pulp**. The small fruits we usually think of as berries, such as strawberries, blueberries and raspberries, are not considered part of the berry family by scientists who study plants. These scientists, called **botanists**, group fruits such as oranges, lemons, and grapefruits together as part of the berry family. These berries are often classed together as citrus fruits.

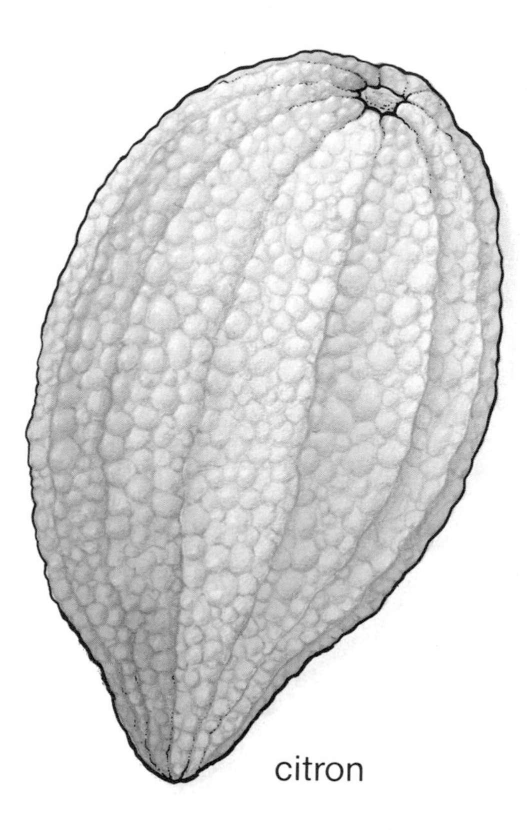

citron

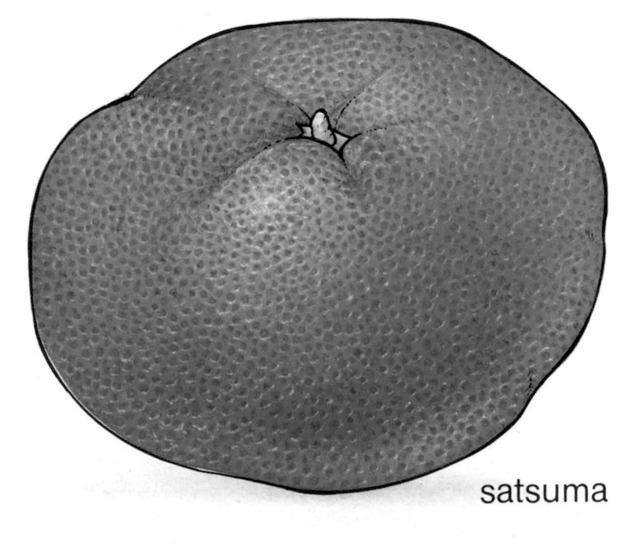

satsuma

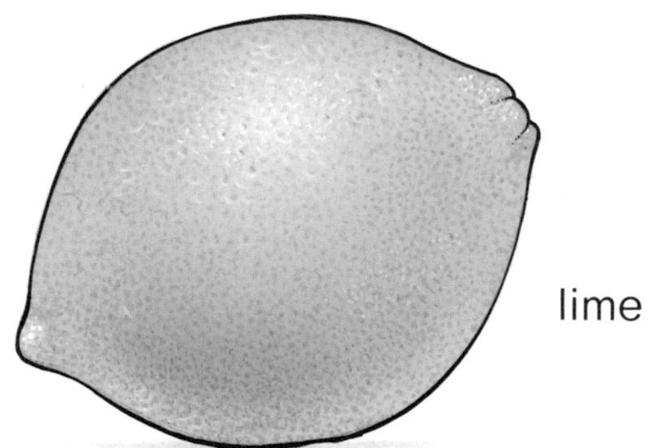

lime

Oranges

Oranges first grew wild in China and southeast Asia. People have grown them since early times. It is believed that Arabs brought oranges to southern Europe in the ninth century, and most European languages get their name for the fruit from the Arabic word *naranj*. Gradually, orange plants were taken to other European countries and eventually to the United States.

There are two main types of orange—sweet and sour. Sweet oranges are one of the most widely grown fruits in the world. We eat sweet oranges raw and use them to make orange juice. The most popular variety of sweet orange in the world is the Valencia. Sour or bitter oranges are mainly grown in Europe. They have bitter juice and rough skins. They are used for making marmalade but are not eaten raw.

Left: A colorful display of oranges on sale at a market in Bolivia

Below: Picking oranges on a fruit farm in Australia

Orange trees thrive in warm climates where there is no frost. Full-grown trees may bear up to 1,000 oranges a year and can produce good crops for as many as 50 to 80 years.

Brazil and the United States are the world's leading producers of oranges. In the United States, most oranges are grown in Florida, but California and Texas also have large orange crops. Other countries that are major growers of oranges include Mexico, Spain, and Israel.

Tangerines, mandarins, kumquats, and satsumas are all closely related to oranges.

Lemons

It is thought that lemons first grew wild in parts of India, Burma, and China. Arabs brought lemons to Europe from India in the early Middle Ages. Christopher Columbus brought the first lemons across the Atlantic on his voyage to Haiti in 1493. The first lemon trees in the United States were probably planted by Spanish missionaries in the 18th century.

Lemon trees need less heat than other citrus fruits, but still require a warm climate. They have been successfully grown outdoors, in sheltered conditions, as far north as England. In cold climates, the lemon tree is often grown with other citrus fruits in an **orangery**. The main lemon-growing regions of the world are Italy, Spain, the United States (primarily California and Arizona), Israel, and North Africa.

Left: Lemons ripening on a tree in the spring sunshine in Spain

Below: This boy is enjoying a glass of lemonade, one of the many products made from lemons.

There are many varieties of lemon. Some are rounded, with rough, pale yellow skins. Others are smooth-skinned and dotted with oil glands. Lemons are not usually eaten by themselves because of their sour taste. However, lemon juice and lemon oil are often used in cooking to add flavor to meat, fish, and some desserts. Lemon oil is also frequently used as a fragrance in perfumes, soaps, and shampoos.

Limes

It is not known where limes came from, but they probably originated in India. Arabs probably brought limes to Europe from India during the Middle Ages. Eventually, explorers brought limes

Harvesting limes in the West Indies. These limes will be used to make juice and marmalade.

A woman sets out a display of limes to sell at a market in Acapulco, Mexico.

to other parts of the world.

Limes grow on small trees that bloom with clusters of scented white flowers. Limes are small, usually about two inches long, and are slightly rounder than lemons. Their color is, of course, lime green.

Limes are grown in many tropical regions, such as the West Indies, the Mediterranean countries, Mexico, Egypt, Israel, Florida, and southern California.

Grapefruit

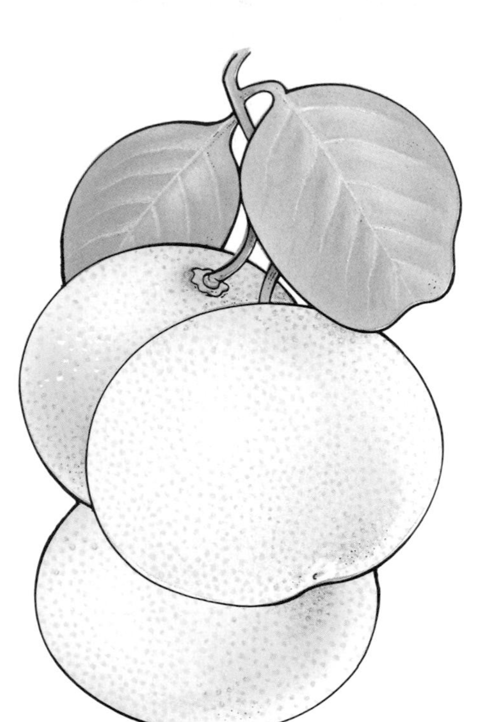

The exact origin of the grapefruit is unknown, but it is probably descended from the larger shaddock or pomelo fruit found in India and southeast Asia.

Grapefruits were given their name because they grow in clusters, like grapes. They grow on evergreen trees that bear fragrant white flowers.

There are many varieties of grapefruit, all with

pink grapefruit

pomelo

ugli fruit

This Israeli man is harvesting grapefruits by hand.

different shapes and thicknesses of skin. Pink grapefruit have a rosy-colored pulp. **Ugli fruit** are a cross between grapefruits, oranges, and tangerines.

The southern United States produce a large part of the world's supply of grapefruits, but the West Indies, South Africa, Israel, Greece, and Spain are some of the other important grapefruit-growing regions of the world.

Citrus fruits for health

Citrus fruits are an important part of a balanced diet. They are rich in vitamins and minerals. Our bodies need only small amounts of vitamins and minerals, but without them we would not be healthy.

Citrus fruits are rich in **vitamin C**. Too little of this vitamin leads to a disease called **scurvy**. At one time, scurvy was common among sailors. When sailors were away at sea for long periods, there was

Citrus juices were given to sailors to prevent scurvy, a disease caused by a lack of vitamin C.

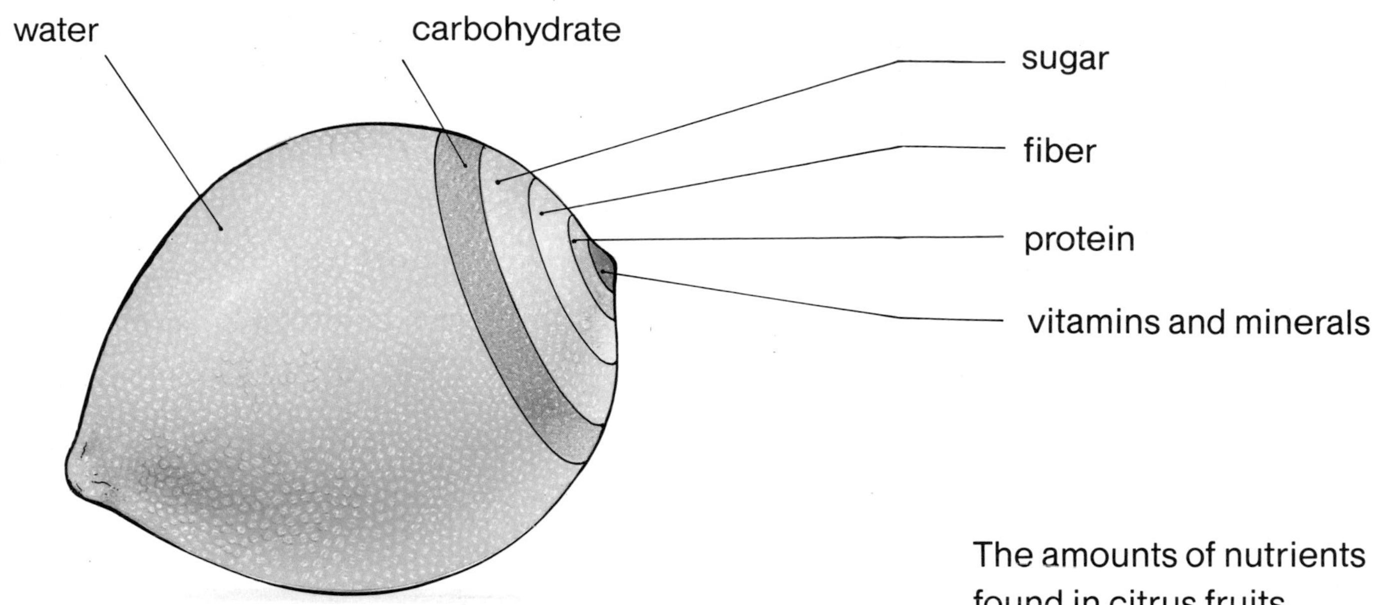

The amounts of nutrients found in citrus fruits

no way for them to get fresh fruits and vegetables to eat, so their diets contained no vitamin C.

By the end of the 16th century, people knew that orange and lemon juice prevented scurvy, but it was not until the 18th century that doctors discovered it was the vitamin C in citrus fruits that cured the disease.

Most people today eat enough foods containing vitamin C and do not have to worry about scurvy.

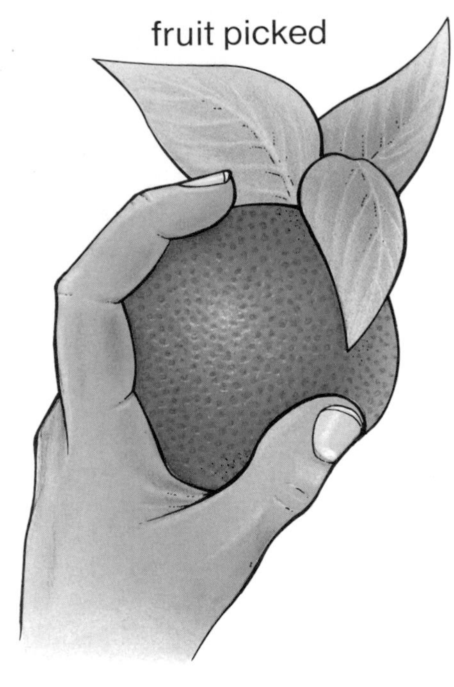

Growing citrus fruits

Citrus fruit trees grow best in warm climates where there is no frost during the year. In colder areas, they can be grown in greenhouses or other protected areas.

The soil in which citrus trees are grown must be **fertile**, so manure or chemical fertilizers are added. These fertilizers provide nutrients that help

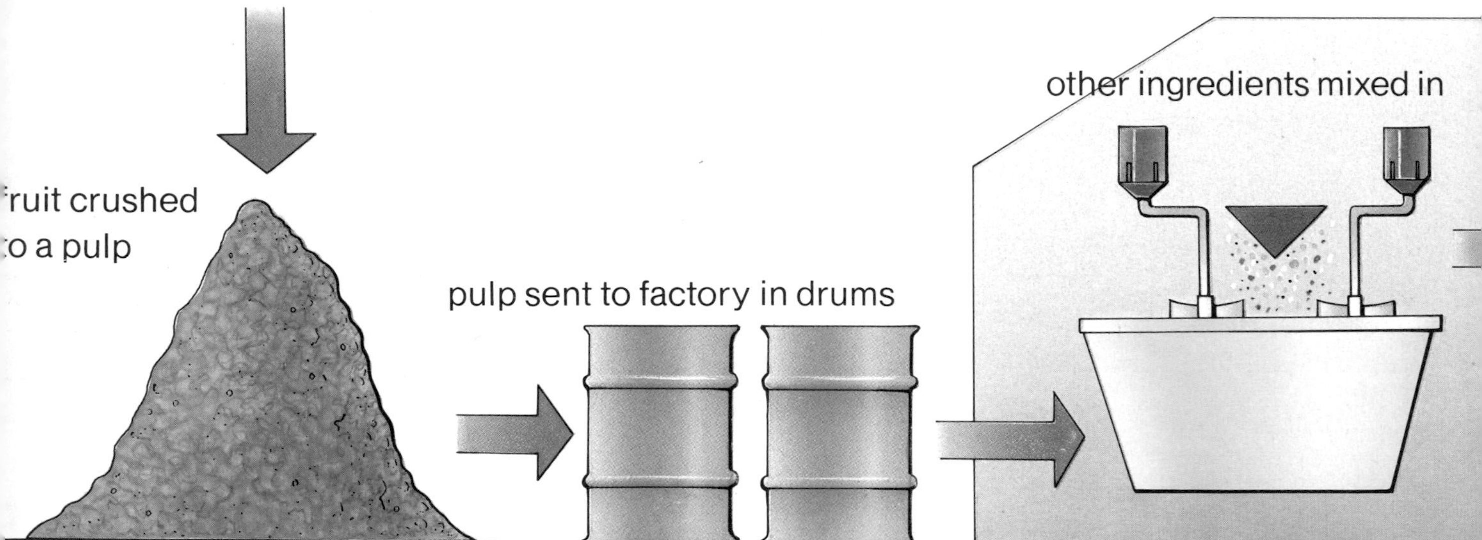

the plants grow well. Some growers use chemical sprays to stop the fruits from falling from the trees too early. Many growers also use **pesticides** and **fungicides** to protect the trees from insects and disease.

Over the years, new varieties of citrus trees have been developed. Some will bear larger fruit; others produce a more abundant crop. Stronger trees that are resistant to disease have also been developed. Some varieties have been specially adapted to suit

This diagram shows how fruit juices are made from citrus fruits.

particular climates.

Some trees that produce good fruit are delicate and easily killed by disease. Through a special process called **grafting**, shoots from these trees may be joined to trees that are especially strong and resistant to disease. This grafting will produce strong trees that bear good fruit. **Dwarf trees**, which need less room than full-size trees, have also been developed in areas where little land is

Citrus fruit trees grow best in warm climates. These are orange groves in Spain.

Spraying orange trees with pesticides on a farm

available for crops.

Citrus fruits that are to be sold fresh are usually picked by hand. They are then shipped to packing houses, where they are sorted according to size and quality, then packed and shipped to markets. Some citrus fruits are picked by machines and shipped to factories to be made into juices and other products.

How we use citrus fruits

Fresh-squeezed oranges can be made into a refreshing, healthy drink.

When they are fresh, citrus fruits have bright, slightly moist skins. Citrus fruits stay fresh for a long time, but they must be used before their skins shrivel. We enjoy drinking the juices of citrus fruits, and some are eaten raw. But their flesh, juice, and peel can also be used in cooking. The fruits may be used as garnishes, too. Drinks such as lemonade and orange drink, and desserts such as sherbet are all made with citrus fruits.

Citrus fruits, because they are rich in vitamin C, are thought by many to help cure the common cold, but vitamin C has never been proved to be a cold remedy.

The extracts of some citrus fruits may be used in cleaning products. For example, lemon and lime oils are often used in household cleaners.

Fruits may be **preserved** to be eaten at a later

time. Tiny organisms called bacteria will eventually cause food to decay, but there are various ways of keeping fruits from decaying. One is to preserve them in sugar. Bacteria do not grow in sugar

Citrus fruits are often used to flavor various foods. They also make delicious marmalades, jams, and jellies.

solutions, so jams, marmalades, preserves, and jellies stay fresh for a long time. Candied fruits are also made with sugar.

Two traditional methods of preserving fruits are canning and drying. Nowadays fruits are also preserved through freezing, since bacteria cannot

Left: Grapefruit is a popular and tasty breakfast food that is full of vitamin C.

Left: Unloading grapefruits at a factory in Israel. These will be peeled, divided into segments, and canned.

Below: Some of our favorite desserts are made with citrus fruits.

survive at very low temperatures. Citrus fruits can also be preserved by being pickled or made into **chutneys**.

These methods allow us to eat fruits that may have been harvested many months before, but preservation can destroy some of the nutrition in the fruits.

How to make a pomander

In the days when the bubonic plague spread across Europe in the Middle Ages, **pomanders** were believed to ward off infection. Today they are used to scent clothes and linens.

You will need:

cloves, a thin-skinned orange that has been dried for at least two weeks in a warm place, a skewer, gauze tape, a ribbon, plastic-headed pins, 1 teaspoon of orris root powder (available in craft stores) if desired, and 1 teaspoon of cinnamon.

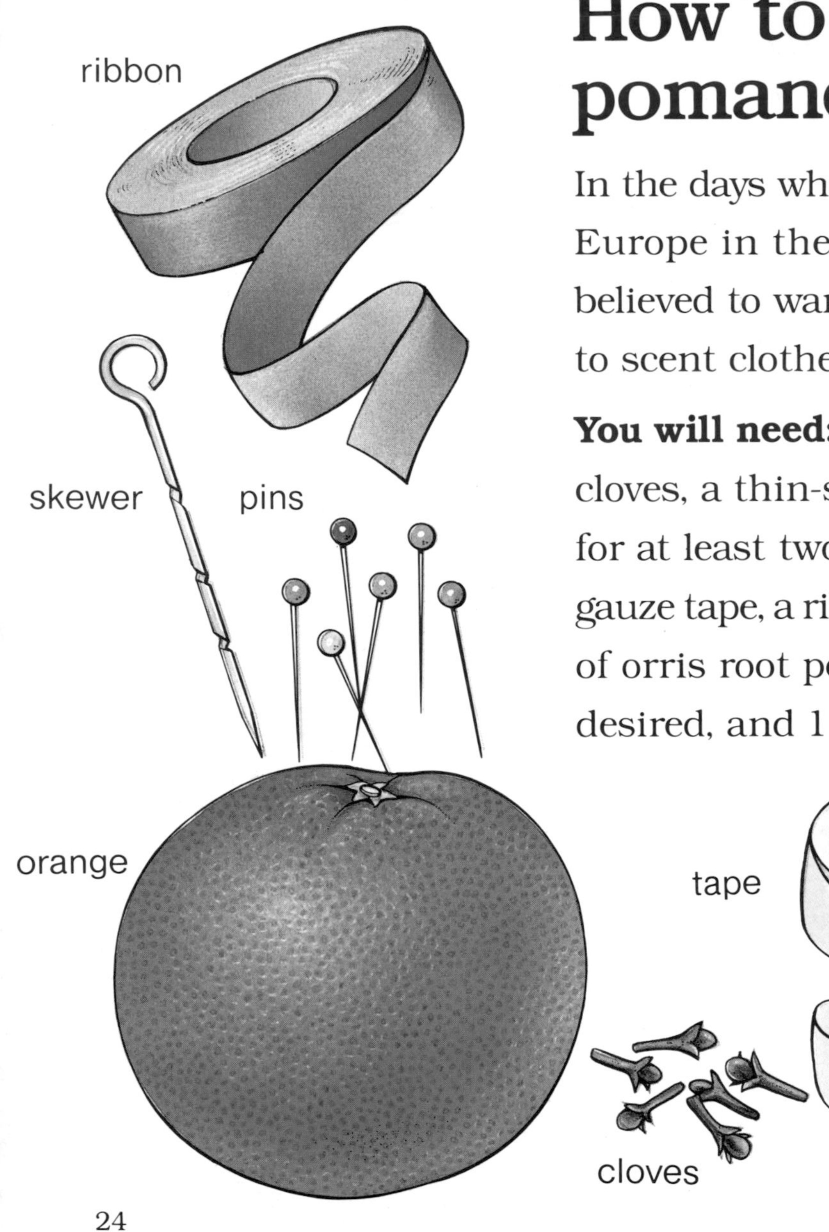

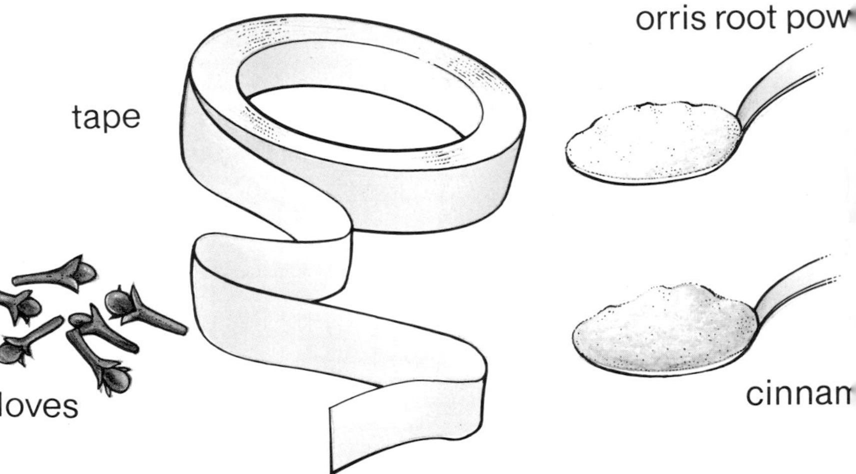

orris root pow

cinnam

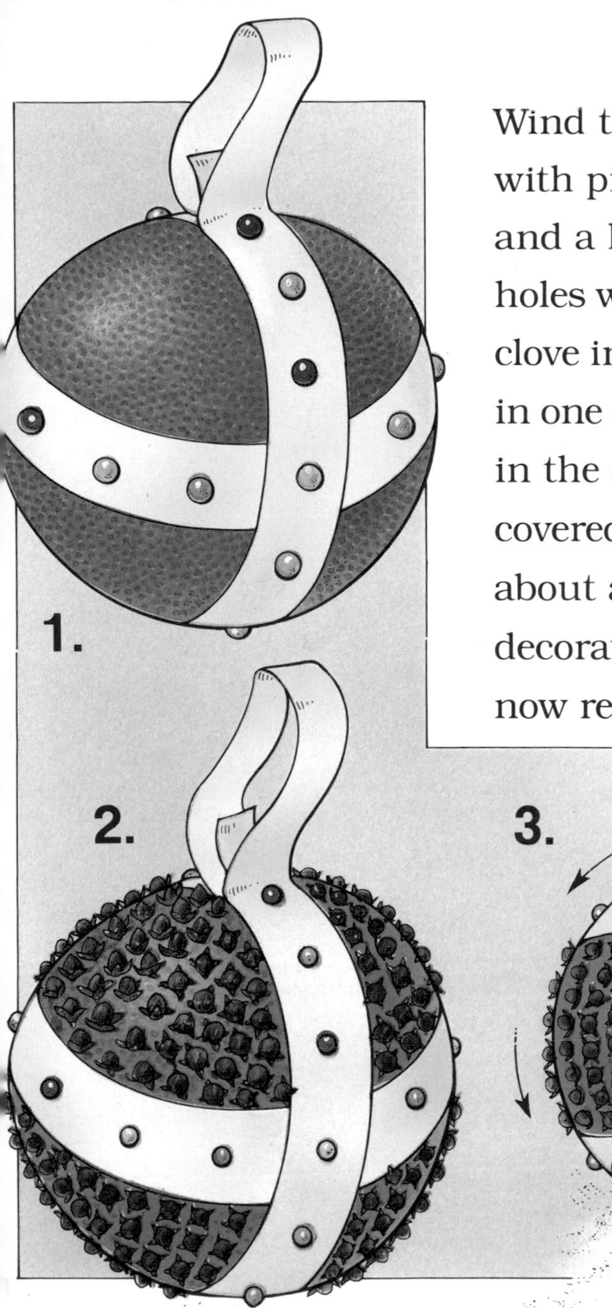

Wind the tape around the orange and secure it with pins. You can make two, four, or six sections and a hanging loop with the tape. Make small holes with the skewer all over each section. Press a clove into each hole. Make sure that you do this all in one day, or mold may develop. Roll the pomander in the orris root powder and cinnamon until it is covered, and hang it in a warm, dry place for about a week. You may replace the tape with a decorative ribbon if you like. The pomander is now ready to use.

Make sure there is an adult nearby when you are cooking.

Lemonade

You will need:

3 lemons
3 cups boiling water
1/4-1/3 cup sugar

1. Wash the lemons well. Grate the peel of two lemons and put in a large bowl.

2. Carefully pour in the boiling water. Add the sugar and stir until it is dissolved. Cover the bowl and leave it to cool.

3. Squeeze the juice from the lemons and add it to the cooled liquid.

4. Strain and chill before serving.

You can make orangeade by using two oranges and one lemon and only 1/4 cup sugar.

Orange rice

You will need:

2 tablespoons butter
¼ cup finely chopped celery
3 tablespoons chopped onion
⅔ cup orange juice
1 tablespoon grated orange peel
a pinch of salt and pepper
⅔ cup long grain rice
1¼ cups water
1 tablespoon chopped parsley

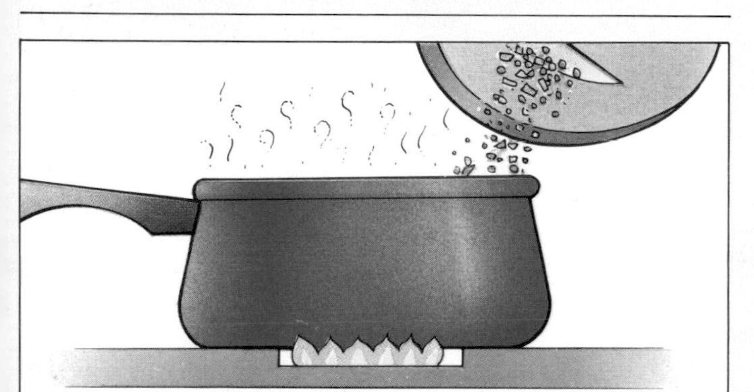

1. Melt the butter in a saucepan. Gently fry the celery and onion until they are tender, but do not brown them.

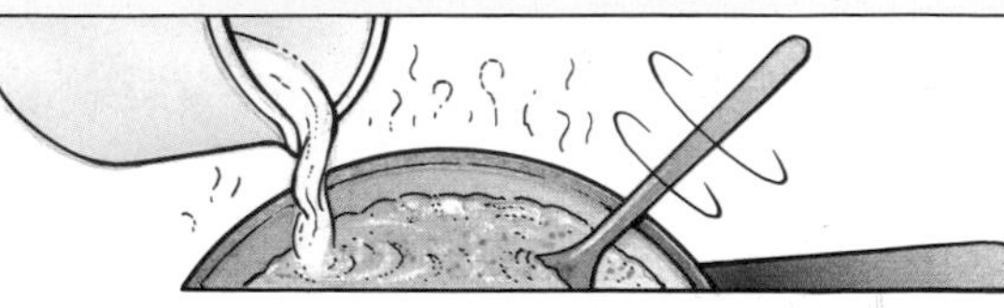

2. Stir in the water, juice, orange peel, and the salt and pepper. Bring to a boil.

3. Rinse the rice and sprinkle it into the pan.

4. Put the lid on the saucepan and cook over low heat for about 20 minutes, until all the liquid is absorbed.

5. Spoon the rice into a serving dish and fluff it up with a fork. Sprinkle the parsley on top.

Orange and lemon chutney

You will need:

2 oranges, peeled
2 lemons, peeled
1 cup water
2 large onions, chopped
1 cup chopped raisins
½ teaspoon ginger
1½ cups sugar
a pinch of salt
½ teaspoon cayenne pepper
2½ cups malt vinegar

1. Chop the peeled oranges and lemons into small pieces and put them into a large saucepan. Add the water.

2. Add the chopped onions, raisins, ginger, and sugar, and simmer on a low heat until the mixture is soft.

3. Add the salt, cayenne pepper, and vinegar, and simmer slowly until thickened. Stir the mixture frequently.

4. Pour into jars and cover immediately. Leave to cool. This delicious chutney can be eaten with salads or spicy foods.

Glossary

berry: any member of a family of fruits that have their seeds in their pulp

botanists: scientists who study plants, including fruits

chutneys: relishes made from fruits and vegetables. Chutneys are often eaten with spicy foods.

dwarf trees: specially produced trees that are small and need less room to grow than full-sized trees

families: groups of fruits that are similar to each other

fertile: able to produce living things

fungicides: chemicals that are used on plants to keep them from getting diseases

grafting: joining parts of two plants together to create a new plant

orangery: a special sheltered area where oranges and other citrus fruits are grown

pesticides: chemicals that are used on plants to kill insects

pomanders: fragrant, clove-studded oranges that are used to scent clothes and linens

preserved: kept from rotting by various means, such as canning and freezing

pulp: the soft, fleshy part of a fruit

scurvy: a disease caused by a lack of vitamin C

ugli fruit: a citrus fruit that is a cross between a grapefruit, a tangerine, and an orange

vitamin C: a vitamin that is found in many fruits and vegetables, especially citrus fruits

Index

Photo Acknowledgments

The photographs in this book were provided by: pp. 7 (left), 11, Hutchison Library; pp. 7 (right), 9 (right), 13, 20, 21, 23, ZEFA; pp. 9 (left), 18, Travel Photo International; p. 10, J. Allan Cash; p. 14, Mary Evans Picture Library; p. 15, Outspan; p. 19, Holt Studios; p. 22, Wayland Picture Library.